Gerry Swan

First published in 2017 by Reed New Holland Publishers
Sydney

Level 1, 178 Fox Valley Road, Wahroonga, NSW 2076, Australia

newhollandpublishers.com

A record of this book is held at the National Library of Australia.

ISBN 978 1 92151 789 1

Managing Director: Fiona Schultz
Publisher and Project Editor: Simon Papps
Designer: Andrew Davies
Production Director: Arlene Gippert
Printed in China

10 9 8 7 6 5 4 3

Keep up with New Holland Publishers:
 NewHollandPublishers
 @newhollandpublishers

CONTENTS

INTRODUCTION

Australia has an undeserved reputation for being infested with dangerous snakes. We are never told that Australia is infested with kangaroos or birds, and the word infested seems limited to use with animals the media want to beat up. The reality of course is that we are not infested with snakes. They are not plentiful in the landscape and are in fact quite secretive, and unless moving are not noticed by most people. Although about 60 per cent of Australian snakes are venomous most are not dangerous to humans.

The aim of this book is to help the reader to identify snakes they may encounter through a combination of photos, description, and information on distribution and habitat — always remember that a particular species will not inhabit every area within its known distribution, and that most are quite particular regarding habitat.

ACKNOWLEDGEMENTS

Thanks to the following people for the use of photographs:

Mike Anthony: pages 47, 53, 62, 66, 67, 94 and 116. Rich Carey, Shutterstock: page 172. Brian Champion: pages 59 and 60. Hal Cogger: pages 171, 172, 173, 174, 175, 176, 177, 178, 179, 180, 181 and 183. Scott Eipper: pages 37, 49, 52, 72, 106, 117, 121, 128, 140, 145, 158 and 169. Adam Elliott: pages 39, 73 and 74. Simon Fearn: pages 14, 17, 50, 90, 110, 125 and 126. Hank Jenkins: pages 55 and 77. Brad Maryan: pages 31, 35, 41, 43, 48, 54, 61, 63, 79, 84, 85, 87, 101, 102, 111, 122, 127, 131, 134, 135, 139, 146, 147, 162, 163 and 164. Ross Sadlier: pages 80, 92, 149, 155, 157 and 167. Craig Searle: page 7. Glenn Shea: pages 30, 33, 95, 96, 120, 133 and 182. Kristina Vackova, Shutterstock: page 171. Chris Watson, Shutterstock: pages 1, 21. Steve Wilson: pages 32, 76 and 78. John Wombey: pages 46, 86, 88, 129 and 153.

All other images by Geoff Swan and Gerry Swan.

Thanks also to Brett Aitcheson, Jim George and Peter Rowe for assistance with photographic material.

5

WHAT ARE SNAKES?

Basically snakes are reptiles that lack limbs, having evolved from some lizard-like ancestor. They are vertebrates that have an elongate limbless body form covered in waterproof scales. To get their internal organs into this tubelike body some have lengthened or become staggered along the body and they have lost one of their lungs. Externally they have a long forked tongue that is constantly flicked in and out, and they do not have eyelids. Because of this they

Snake eating a Ring-tailed possum.

cannot shut their eyes and consequently have a constant unblinking gaze. No snake has an external ear opening.

All snakes are carnivores, hunting down and eating other animals. They cannot break up a food item so it must be swallowed whole. To do this the jaw is loosened to expand the mouth and so consume prey that is much greater in diameter than the snake's body.

TELLING SNAKES AND LIZARDS APART

Legless lizard.

A reptile with limbs will be a lizard as no snake has legs, although pythons have a very small spur either side of the vent that are vestigial hind legs. The majority of lizards have obvious legs, but in some cases these are very reduced and difficult to see without handling the animal.

Snakes all have forked tongues that are constantly flicked in and out of the mouth. No legless or near-legless lizard has a forked tongue, instead they all have broad fleshy tongues. No snake has an

Snake.

external ear-opening but most lizards do, although it may be quite small in some species.

The belly scales in snakes are in a single row whereas lizards have paired or irregular belly scales. Snakes have relatively short tails whereas the tail of a lizard is quite long and may be longer than the body. In legless lizards the tail is easily cast off but snakes do not cast off their tails.

9

SEEING SNAKES

Always remember that snakes are protected and that it is illegal to disturb them in any way.

Snakes do not like hot sunlight, so looking for them during the middle of a hot day will be unrewarding. They are more likely to be out basking or foraging early in the morning or late afternoon and at night. In warm weather walking through the bush or along tracks before 10am or after 4pm could get some results, although many motionless snakes are difficult to see. Searching at night along tracks with a torch can be interesting but remember to take extra batteries.

They are often wary and well camouflaged so it will usually be the larger more active snake that you see. To find smaller more secretive snakes, looking under building rubbish, logs or rocks can produce results, particularly in cooler weather. Remember to replace

A partially concealed snake.

rocks and logs in exactly the position you found them.

One of the best methods of seeing snakes is by spotlighting from a car. This involves travelling along a quiet bushland road at night using the vehicle headlights. This is best done with two persons, driving at a speed of about 50km per hour. Watch out for other vehicles behind you and pull over to let them pass. When you spot something – usually an eye reflection – pull off the road and use your torches to check the snake. Always remember to carry a snake bite kit and know what to do in the event of a bite.

Searching around a derelict building.

11

IDENTIFYING SNAKES

There is no clear method of distinguishing non-venomous snakes from venomous snakes. In many cases colour is not reliable as there is such variation in individuals of the same species when it comes to colour and pattern. In some cases juveniles can be very different from adults of the same species. Not all brown-coloured snakes are from the brown snake family, and not all individuals from the brown snake family are brown – they can range from black to pale tan. Similarly not all individuals from the black snake family are black.

The only sure way to identify many snakes is by counting scales across the middle of the body and the belly. These can differ between similar-looking snakes. However it is beyond the scope of this book to provide such details for identification.

In this book the 'length' measurements in the species accounts give the average and maximum lengths of an adult of that species.

There are many books available on Australian snakes, some of which are listed in the Further Reading section of this book. Some cover the whole of Australia, others cover a particular state or region. Get those that are relevant to your area and familiarise yourself with the snakes that are found there. Also worthwhile to spend time at the local reptile park or zoo and see some of the local species.

Where possible take photos so that you can check these against pictures in field guides or on the web.

What SNAKE is That? Introducing Australia's Snakes — Gerry Swan and Steve Wilson

Australian SNAKES A NATURAL HISTORY — WHITLEY AWARD WINNER

REPTILES OF AUSTRALIA — Steve Wilson and Gerry Swan — FOURTH EDITION

snakes OF WESTERN QUEENSLAND A Field Guide

SNAKES OF WESTERN AUSTRALIA — L.A. Smith and R.E. Johnstone — WA Museum

RICK SHINE

A FIELD GUIDE TO REPTILES OF NEW SOUTH WALES SECOND EDITION — Gerry Swan, Glenn Shea and Ross Sadlier

SNAKES OF TASMANIA — SIMON FEARN

REPTILES WORLD

Steve's Guide to snakes of the ... and ... NSW

SNAKES OF ... QUEENSLAND

SNAKES of ... and ...

SNAKES SOUTH-EAST AUSTRALIA — Australian Reptile Park's Guide to Snakes

Common Snakes of the Katherine Area

WHAT DO SNAKES EAT?

All the Australian snakes are carnivorous, that is they catch and eat other animals. They do not eat any vegetable matter and because they cannot dismember their prey it is eaten whole. Blind snakes eat the eggs, larvae and pupae of termites and ants, while most other snakes catch and eat larger, faster-moving prey. What a snake eats depends very much on its size, where it lives and what species it is. Some are generalists which eat whatever they come across, be it frogs, lizards, other snakes, birds or mammals. Others are specialists, such as the bandy bandys that eat blind snakes, or the shovel-nosed snakes that eat reptile eggs.

Newly born pythons seek out and eat small prey such as lizards, but as they get bigger they hunt larger food items such as rats, possums and rabbits. Many are known for their ability to eat prey much larger than themselves because of an elastic skin and a loosely articulated lower jaw, giving the snake a huge gape. It can work each side of the jaw forward separately, 'walking' the mouth over the prey. Some snakes actively hunt for their food and chase it down while others rely on ambush and will stay in one place for days waiting for their meal to come past.

Above: Tiger Snake eating a rat.

Below: Keelback attempting to eat a frog.

REPRODUCTION IN SNAKES

Above: Python incubating eggs.   *Opposite: Tiger Snake with her 51 young.*

Snakes are either egg-laying or live-bearing. Generally those that have live young inhabit the cooler regions where incubating eggs is a more risky option, while egg-laying species are more likely to occur in temperate and tropical regions.

Snake eggs are leathery and soft-shelled, unlike those of birds which are hard-shelled. The female snake will deposit her eggs in a warm and humid spot under a rock, in a log or burrow or in a crevice. She then leaves them and plays no further part in their development. The eggs incubate for a period of 40–70 days, then the fully developed young will cut the egg with a special egg tooth and emerge.

Pythons are the exception and the female will coil around the eggs to protect them and keep them warm. She maintains this care until the young emerge from the eggs at which time she departs and they are left to fend for themselves.

In live-bearing species the female will find a secluded spot in which to give birth to the litter of young. They are each in a transparent membrane that they break out of as soon as they are born. They then disperse in different directions and are quite independent of the parent.

17

SHEDDING

Milky eye prior to shedding.

Shedding or sloughing of the skin occurs in snakes several times a year, more often in younger snakes. The outer part of the skin of a snake is made up of keratin, which is similar to our fingernails. This is what forms a snake's scales, which are joined by a thinner material to form one continuous layer. This skin is unable to expand indefinitely as the snake grows and is therefore discarded.

When shedding is due to commence the old skin becomes cloudy

Working the skin over the snout.

or milky, particularly on the eyes. This condition disappears after a few days and the snake then starts the sloughing process by rubbing the nose and jaws against some rough material such as bark or rock. This causes the old skin to peel back and the snake literally slides out of the old skin, which is discarded inside out.

Usually the old skin comes away in one piece and is a lot longer than the snake itself because it has stretched as it peels off.

19

MYTHS ABOUT SNAKES

Are snakes wet and slimy? No. Snakes are quite dry to the touch unless they have just emerged from water. Probably the glossy nature of the skin of many snakes has given rise to this myth or that people have mistaken slimy eels for snakes.

Does the forked tongue of a snake sting in any way? No. The forked tongue is a tasting and smelling organ and cannot inflict any sting or bite.

Do snakes only die at sunset? No. Snakes will die at any time of the day or night. However the body of a dead snake may exhibit movements because of the nervous system.

Do snakes drink milk from cows? No. It is very doubtful that any cow would take kindly to have her teat grabbed by a mouth full of sharp teeth. In earlier times primitive facilities meant that mice and rats would be present around livestock and these are much more

The 'stinging' tongue.

A 'venomous' Carpet Python.

interesting to snakes. Snakes are not known milk drinkers. In fact they show a distinct dislike for milk.

Tales of taipans breeding with carpet pythons to produce venomous pythons are often heard in north Queensland pubs. This is no more true than sheep being crossed with pigs to produce something for a mixed grill. The other fantasy often spoken about is snakes keeping pace or outrunning horses at full gallop. Snakes are only capable of short bursts of speed and then not much more than 20km per hour.

Do hoop snakes exist? No. Snakes that can seize their tail in the mouth and roll along the ground like a hoop are a figment of the imagination.

KEEPING SNAKES AS PETS

If you are considering keeping a snake as a pet, think about it carefully. Many are long-lived and have quite critical environmental needs. They are not domesticated animals and are not affectionate in the manner of cats or dogs. However, they do get used to the presence of people, are suitable for keeping indoors, don't take up much space, are not noisy and don't need food or exercise on a daily basis.

These days suitable indoor cages can be purchased from pet shops and are set up with heating and lighting. While most snakes are best kept in indoor cages, some of the larger pythons can be held in outside enclosures.

Snakes are protected animals throughout Australia so you must find out the requirements for keeping snakes in your state or territory from the local fauna authorities. You cannot take snakes from the wild – they must be obtained from pet shops or reptile-keepers who have bred them.

But before getting the snake you should obtain a book dealing with the keeping of snakes. There are several available that will explain food requirements and frequency, caging, heating and lighting as well as covering possible problems. It is very important to read up, talk to keepers, get on the web and find out as much as possible before organising a cage or obtaining the snake.

Never release captive snakes into the bush. Contact fauna authorities regarding the disposal of your unwanted pet snake.

Outdoor enclosure for larger pythons.

SNAKE BITE

REMEMBER: not all snakes are venomous and few are dangerously so. Very few snake bites are fatal and often no venom is injected. Symptoms vary but usually consist of a headache, nausea, weakness and sweating. Pain and swelling of the lymph nodes is also common.

If you are looking for snakes or just walking in the bush, you should carry a snake bite first aid kit and know how to use it.

In the event of a bite first aid must be applied quickly.

DO

- Keep the victim calm and still.
- Immediately apply a broad bandage or cloth strip over the bite site, applying the same pressure as for a sprain.
- Apply the bandage over as much of the limb as possible.
- Immobilise the bitten limb with a splint.
- Bring transport to the victim if possible.
- Get medical aid to the victim as soon as possible.
- Maintain a watch on the victim. If breathing stops initiate appropriate resuscitation.

DO NOT

- Waste time trying to catch or kill the snake. Assist the victim instead.
- Cut, suck or wash the bite.
- Use tourniquets or let the victim drink stimulants such as coffee or alcohol.

The site of bites from many Australian elapids often show virtually no visible marks and little pain. By contrast the bites of some overseas species are painful and result in obvious skin lesions.

Below is a photo of a bite on the little finger from a Coastal Taipan. The snake in question is shown on page 130. The owner, an experienced snake handler, immediately applied a broad pressure bandage that was always to hand, and called 000. After receiving antivenom he spent two days at the hospital in intensive care before being discharged.

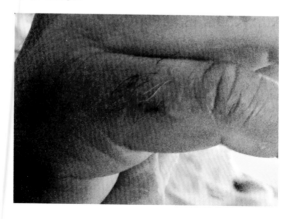

Just two pinpricks.

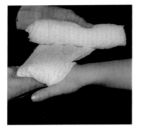

Start bandaging at the bite

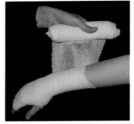

Continue bandaging up the limb

Bandage the whole limb

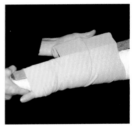

Fix a splint to the limb

Blind Snakes or Worm Snakes
Family Typhlopidae

Burrowers that live mostly underground, they somewhat resemble earthworms and are unlikely to be confused with any other snake. Scales uniform in size around body, close-fitting and smooth; this probably enables them to resist the bites and stings of ants. Eyes reduced to dark spots on head. Tail short and in most species terminates in spur or spine. Small mouth underneath overhanging snout.

Feed exclusively on ant eggs, larvae and pupae, as well as termites. Usually found beneath well-embedded rocks, inside ant nests or termite mounds, and within rotten tree stumps. Sometimes seen on surface at night, particularly if conditions wet. When handled can release offensive smell from anal glands.

Dark-spined Blind Snake *Anilios bicolor*

LENGTH/ID: Average 25cm, maximum 50cm. A moderately robust snake that is grey-brown to purplish-brown above and white below. Border between dark upper and pale lower regions is prominent and jagged. Snout rounded when viewed from above and angular in profile. Spine at end of tail is dark and prominent.

RANGE: Southern Australia from inland NSW and western Vic across to Kalgoorlie area of WA.

HABITAT/HABITS: Usually found in mallee spinifex and semi-arid shrublands.

NOT VENOMOUS, probably incapable of biting a person.

Prong-snouted Blind Snake *Anilios bituberculatus*

LENGTH/ID: Average 30cm, maximum 45cm. Moderate in size and with a slender build. Reddish-brown to blackish above and creamy-white to pink below. Snout distinctly trilobed when viewed from above and slightly angular in profile.

RANGE: Southern WA and SA, northern Vic, NSW west of the dividing range and southern Qld.

HABITAT/HABITS: Open woodland, hummock grasses and shrubland in arid and semi-arid areas.

NOT VENOMOUS, probably incapable of biting a person.

Desert Blind Snake *Anilios endoterus*

LENGTH/ID: Average 30cm, maximum 40cm. Grey-brown to red-brown above with paler snout. White below. Border between the upper and lower colours is prominent and jagged. Snout is weakly trilobed when viewed from above and angular in profile.

RANGE: Central Australia from eastern WA to western NSW and Qld.

HABITAT/HABITS: Red sandy soils in hummock grasslands and shrublands in semi-arid and arid areas.

NOT VENOMOUS, probably incapable of biting a person.

Northern Beaked Blind Snake *Anilios grypus*

LENGTH/ID: Average 35cm, maximum 45cm. A very slender blind snake that is pink to pale red-brown above and cream below. Snout pale and head slightly darker than body. Tail black. The snout is angular when viewed from above and beak-like in profile.

RANGE: Scattered distribution across the interior of northern Australia from coastal WA to central Qld.

HABITAT/HABITS: Generally found in dry habitats with open woodland or grasslands.

NOT VENOMOUS, probably incapable of biting a person.

Robust Blind Snake *Anilios ligatus*

LENGTH/ID: Average 30cm, maximum 50cm. A very thick-bodied blind snake that is reddish-brown to dark brown above, with a pale snout. Cream below, and border between upper and lower colours is distinct and sharply demarcated. Snout appears rounded when viewed from above and in profile.

RANGE: Eastern Qld and northern NSW with a disjunct population from north-western Qld to Kimberley region of WA.

HABITAT/HABITS: Open woodlands. Known to feed on eggs and larvae of large bulldog ants.

NOT VENOMOUS, probably incapable of biting a person.

Blackish Blind Snake *Anilios nigrescens*

LENGTH/ID: Average 40cm, maximum 65cm. A moderately built snake that is grey, pink-brown to blackish above and cream below. Has a dark patch either side of vent. Snout pale and is rounded when viewed from above and in profile.

RANGE: Throughout eastern Australia from Vic, through NSW, to southern Qld.

HABITAT/HABITS: Wide range of habitats including open woodlands, dry and wet sclerophyll forests and rainforests. Frequently occurs where there are well-embedded rocks or fallen timber. Often in suburban areas.

NOT VENOMOUS, probably incapable of biting a person.

Woodland Blind Snake *Anilios proximus*

LENGTH/ID: Average 40cm, maximum 70cm. A large robust blind snake that is grey-brown to dark brown above and creamy-white below. Sometimes has a dark patch on either side of vent. Snout is bluntly trilobed when viewed from above and angular in profile.

RANGE: Eastern Australia from Vic, through NSW, to north-eastern Qld.

HABITAT/HABITS: Found in a variety of habitats from shrublands and open woodlands to wet sclerophyll forests.

NOT VENOMOUS, probably not capable of biting a person.

Southern Beaked Blind Snake *Anilios waitii*

LENGTH/ID: Average 35cm, maximum 60cm. A very slender blind snake that is yellowish-brown to dark purplish-brown above and pinkish-white below. Snout strongly trilobed when viewed from above and angular in profile with a distinctly beaked appearance.

RANGE: Endemic to WA where it occurs in southern and south-eastern regions.

HABITAT/HABITS: Hummock grasslands and open forest, usually on more sandy soils.

NOT VENOMOUS, probably not capable of biting a person.

Brown-snouted Blind Snake *Anilios wiedii*

LENGTH/ID: Average 23cm, maximum 30cm. A small and slender blind snake. Pink-brown to reddish-brown above and cream below. The border between the darker upper and paler lower colours is distinct and sharp-edged. Snout rounded when viewed from above and in profile.

RANGE: Along north-eastern interior of NSW, through south-eastern Qld to about Mackay.

HABITAT/HABITS: Occurs in shrublands and open woodlands.

NOT VENOMOUS, probably not capable of biting a person.

Flowerpot Snake *Indotyphlops braminus*

LENGTH/ID: Average 12cm, maximum 17cm. Slender, very small blind snake. Purplish-brown to blackish above, with paler snout. Pale brown below. Numerous very small tubercles on head scales. Snout rounded when viewed from above and rounded in profile.

RANGE: North-west coast of WA, Darwin area in NT, Torres Strait islands, and coastal Qld south to Brisbane.

HABITAT/HABITS: Accidentally introduced from Asia. Population consists entirely of females that reproduce without need for males (parthenogenesis). Only one individual is required to commence a new colony. Usually found in gardens.

NOT VENOMOUS, probably not capable of biting a person.

Pythons   Family Pythonidae

Pythons are constrictors that strike hard at their prey and quickly coil around it, squeezing tightly to prevent it from breathing. They have numerous, long backward-curved teeth and while they are not venomous, all are capable of inflicting a painful bite if molested.

All pythons are egg-layers and the female will coil around the clutch of eggs to protect them until they hatch. To assist incubation she will sometimes raise her body temperature by basking or shivering.

Most pythons have heat-sensory pits along the lips that detect heat from warm-blooded animals and assist in tracking prey. The Woma and Black-headed Python lack these pits but, as their food is mainly other reptiles, these are unnecessary.

Children's Python *Antaresia childreni*

LENGTH/ID: Average 80cm, maximum 1.1m. Pale brown to reddish-brown with pattern of numerous darker smooth-edged blotches that are prominent in hatchlings but become indistinct or absent with age. Paler on lower sides to creamy-white on belly.

RANGE: From Kimberley region in WA across northern Australia to north-western Qld.

HABITAT/HABITS: Rocky slopes in woodland and open forest where it shelters in animal burrows, hollow timber, rock crevices and caves. Nocturnal and mainly terrestrial, feeding on small lizards, mammals and birds.

NOT VENOMOUS but can still bite.

Spotted Python *Antaresia maculosa*

LENGTH/ID: Average 85cm, maximum 1.25m. Cream to pale brown with numerous prominent ragged-edged darker blotches that often coalesce to form wavy lines along back, usually on fore body or tail. Belly whitish.

RANGE: North-eastern NSW to tip of Cape York Peninsula, generally east of Great Dividing Range.

HABITAT/HABITS: Nocturnal. Woodland, dry sclerophyll forest and rainforest edges, often where there are extensive rock outcrops. Shelters in caves, fallen timber or animal burrows. Feeds on lizards and small mammals, including bats that are taken as they enter or exit caves.

NOT VENOMOUS but can still bite.

Pygmy Python *Antaresia perthensis*

LENGTH/ID: World's smallest python. Average 55cm, maximum 68cm. Pale reddish-brown with small darker brown blotches that become indistinct with age. Belly white to cream.

RANGE: Endemic to WA. Found in the Pilbara region and adjacent areas.

HABITAT/HABITS: Rocky hills with spinifex and shrubs. Also in large termite mounds that are a feature of some areas; sometimes called the Anthill Python because it inhabits these. Feeds on small lizards and mammals.

NOT VENOMOUS but can still bite.

Stimson's Python *Antaresia stimsoni orientalis*

LENGTH/ID: Average 75cm, maximum 1.2m. Cream to pale brown with darker brown blotches that tend to be elongated transversely, with prominent pale stripe along side of body anteriorly. Belly whitish.

RANGE: From eastern half of WA through NT and SA to Qld and western NSW.

HABITAT/HABITS: Inhabits rock crevices, tree hollows, animal burrows and termite mounds across wide range of habitats. Typically found in rocky ridges and outcrops, grasslands and spinifex, and open forest. Food includes lizards and small mammals, particularly bats.

NOT VENOMOUS but can still bite.

Stimson's Python *Antaresia stimsoni stimsoni*

LENGTH/ID: Average 90cm, maximum 1.2m. Fawn to pale brown with prominent reddish-brown to dark brown blotches that are smooth-edged, circular or elongated. Pale lateral stripe on anterior part of body. Belly whitish.

RANGE: This subspecies confined to western WA, from the Kimberley south to Perth. Replaced by *A. s. orientalis* in eastern interior of WA.

HABITAT/HABITS: Nocturnal but occasionally seen basking during day. Rock outcrops and escarpments as well as sand plains and woodlands. Shelters in rock crevices, animal burrows and hollow trees and under exfoliated rock.

NOT VENOMOUS but can still bite.

Black-headed Python *Aspidites melanocephalus*

LENGTH/ID: Average 1.75m, maximum 2.5m. Head and neck glossy jet black. Body cream to pale brown with numerous reddish-brown to blackish irregular crossbands. Belly yellowish, often with darker blotches.

RANGE: North West Cape, WA, north to Kimberley, across much of NT and northern Qld south to about Mitchell.

HABITAT/HABITS: Terrestrial but climbs in shrubs. Low rocky ranges, shrublands, grasslands and woodlands. Shelters in animal burrows, hollow timber, rock crevices and soil cracks. Lacks heat-sensory pits found on lip scales of most pythons. Feeds on snakes, including venomous species, and lizards such as goannas and dragons.

NOT VENOMOUS but capable of inflicting painful and messy bite.

Woma *Aspidites ramsayi*

LENGTH/ID: Average 1.5m, maximum 2.7m. Body yellowish to pale brown with many reddish-brown bands. Head yellow to orange with black patch above eye on juvenile, which is retained in adults in eastern populations. Belly cream to yellow with darker blotches.

RANGE: South-western WA, where in serious decline. North-west coast of WA across southern NT, northern SA, far western NSW to eastern Qld.

HABITAT/HABITS: Terrestrial and nocturnal. Arid and semi-arid areas in grasslands and woodlands. Shelters in logs and burrows; digs own burrow in sandy areas. Has no heat sensory pits on lips. Feeds on snakes and lizards.

NOT VENOMOUS but capable of inflicting painful and messy bite.

Water Python *Liasis fuscus*

LENGTH/ID: Average 2m, maximum 3m. Long head distinct from neck. Glossy olive-brown to blackish-brown, yellowish on lower sides with yellow belly. Lips and throat creamy-white.

RANGE: Northern WA from about Broome, across northern NT and northern Qld including Cape York Peninsula, and down east coast to about St Lawrence.

HABITAT/HABITS: Semi-aquatic and nocturnal. Floodplains, billabongs and swamps, or woodlands and forests adjacent to permanent water. Shelters in vegetation, logs or animal burrows and takes to water if alarmed. Feeds on rats and other mammals, waterbirds and even young Freshwater Crocodiles.

NOT VENOMOUS but capable of inflicting painful and messy bite.

Olive Python *Liasis olivaceus olivaceus*

LENGTH/ID: Average 3m, maximum 4.5m. Yellowish-brown to dark brown, paler on lower sides with no pattern. Belly creamy-white. Head long and narrow with cream lips.

RANGE: From Kimberley region in WA through northern NT to north-western Qld.

HABITAT/HABITS: A terrestrial and rock-inhabiting species that occurs in tropical and subtropical areas. Found in woodlands, shrublands and riverine areas, particularly around rocky outcrops, caves and rocky pools. Preys on birds, reptiles and mammals.

NOT VENOMOUS but capable of inflicting painful and messy bite.

Olive Python *Liasis olivaceus barroni*

LENGTH/ID: Average 3m, maximum 6.5m. Yellowish-brown to dark brown, paler on lower sides with no pattern. Belly creamy-white. Head long and narrow with cream lips. It is not possible to distinguish between the two subspecies of Olive Python on the basis of colour.

RANGE: This subspecies endemic to WA and found only in Pilbara and northern Ashburton region.

HABITAT/HABITS: Terrestrial and nocturnal. Rocky gorges, around deep pools and watercourses in the ranges. Mammals and birds taken by ambush.

NOT VENOMOUS but capable of inflicting painful and messy bite.

Rough-scaled Python *Morelia carinata*

LENGTH/ID: From few specimens caught average length about 2m. Slender, large-eyed snake with keeled scales that are unique among pythons. Tan in colour with numerous large dark brown to reddish-brown blotches and streaks that are transversely elongated. Belly whitish with pale brown markings.

RANGE: Known only from north Kimberley region in WA, around Hunter and Mitchell Rivers.

HABITAT/HABITS: Arboreal and rock-dwelling, inhabiting monsoon forest in rocky gorges and escarpments.

NOT VENOMOUS but capable of inflicting painful and messy bite.

Western Carpet Python *Morelia spilota imbricata*

LENGTH/ID: Average 1.7m, maximum 2.5m. Greenish-brown to blackish-brown with dark-edged pale cream or brown blotches. These are transversely elongated and form longitudinal lines on anterior flanks. Belly cream, sometimes with black blotches.

RANGE: South-western WA north to about Geraldton and east to Eyre. Also on offshore islands, including Garden Island at Perth.

HABITAT/HABITS: Forested areas and shrublands, particularly around rocky outcrops. Shelters in rock crevices, burrows and logs. Generally nocturnal but will bask during day. Preys on mammals, birds and lizards.

NOT VENOMOUS but capable of inflicting painful and messy bite.

Eastern Carpet Python *Morelia spilota mcdowelli*

LENGTH/ID: Average 2.4m, maximum 4m. Pale to dark brown above, with paler blotches and bands which are darker-edged, and may fuse in varying irregular combinations, often producing a pale lateral stripe. Belly cream to yellow, sometimes with dark blotches.

RANGE: Eastern coast from Cape York Peninsula, Qld, to Coffs Harbour region, NSW.

HABITAT/HABITS: Timbered and rocky areas. Frequently enters buildings on edges of urban areas. Nocturnal and arboreal. Ambush predator feeding on mammals and birds.

NOT VENOMOUS but capable of inflicting painful and messy bite.

Murray/Darling Carpet Python
Morelia spilota metcalfei

LENGTH/ID: Average 1.6m, maximum 3m. Pale to dark grey above, often with reddish-brown flush. Series of roughly paired, pale grey round blotches, which may sometimes be joined, runs down back. Blotches on side join to form longitudinal stripe. Belly cream to grey.

RANGE: Central and western Qld, central and western NSW, northern Vic and eastern SA.

HABITAT/HABITS: Dry woodland, rocky outcrops and large trees along watercourses. Nocturnal, sheltering by day in rabbit warrens, tree hollows and rock crevices.

NOT VENOMOUS but capable of inflicting painful and messy bite. Normally placid but can bite.

Diamond Python *Morelia spilota spilota*

Juvenile.

LENGTH/ID: Average 2m, maximum 2.5m. Black to dark olive with cream or yellow spot on most scales. Clusters of yellow scales ringed by black scales in roughly diamond-shaped patches, scattered over body. Belly cream to white with black marbling.

RANGE: East coast from mid-northern NSW to north-eastern tip of Vic. Most southerly occurring python in world.

HABITAT/HABITS: Rainforest, sclerophyll forest and rocky country where shelters in rock crevices, tree hollows and sometimes buildings. Active day and night, preying on mammals and birds.

NOT VENOMOUS but capable of inflicting painful and messy bite.

Top End Carpet Python *Morelia spilota variegata*

LENGTH/ID: Average 1.6m, maximum 2m. Reddish-brown to blackish-brown with ragged dark-edged pale bands along length of body. Belly cream.

RANGE: From Kimberleys in north-western WA, across northern NT and in north-western Qld.

HABITAT/HABITS: Rocky escarpments, sclerophyll forests, humid woodland and monsoon forests. Also urban areas adjoining forests. Arboreal, sheltering mainly in tree hollows and rocky crevices. Nocturnal but may be seen basking during day. Adults feed on mammals and birds, juveniles will take lizards.

NOT VENOMOUS but capable of inflicting painful and messy bite.

Centralian Carpet Python *Morelia bredli*

LENGTH/ID: Average 2.7m, maximum 3m. Reddish-brown or orange to dark brown with numerous irregular dark-edged cream blotches and transverse bars. Belly white to yellowish with darker grey variegations.

RANGE: Central Australia in southern areas of NT.

HABITAT/HABITS: Dry shrublands or woodlands, often along dry tree-lined watercourses or in rocky gorges. Usually shelters in rabbit warrens, tree hollows, caves or deep crevices. Sometimes found in houses in Alice Springs. Nocturnal and preys upon mammals and birds.

NOT VENOMOUS but capable of inflicting painful and messy bite.

Jungle Carpet Python *Morelia spilota cheynei*

LENGTH/ID: Average 1.4m, maximum 2.3m. Dark brown to black with irregular cream to vivid yellow blotches, often forming transverse bars. Occasionally these become longitudinal lines on fore part of body. Belly cream to yellow with black spots or flecks.

RANGE: Confined to wet tropics of Qld, centred on Atherton Tableland and adjacent areas.

HABITAT/HABITS: Mainly along watercourses draining Atherton Tableland, in rainforests, sclerophyll forests and humid woodlands. Arboreal and nocturnal but will bask during day. Feeds on mammals and birds.

NOT VENOMOUS but capable of inflicting painful and messy bite.

Green Python *Morelia viridis*

LENGTH/ID: Average 1.5m, maximum 2.2m. Adult lime-green with more or less continuous row of white or sometimes yellow scales down vertebral line of back. Usually a few scattered white scales on sides. Short very pale blue transverse bars along vertebral line. Belly cream to yellow.

At birth, hatchlings are bright yellow to orange with purplish-brown blotches and streaks along body. A similarly coloured streak runs from the nostril through the eye to the neck.

Yellow juvenile.

Red juvenile.

RANGE: The Green Python is confined to two areas of rainforest in the Iron and McIlraith Ranges on north-east Cape York Peninsula.

HABITAT/HABITS: Arboreal, found in rainforest and vine thickets. During day rests on branch or vine in distinctive coiled loop with head in centre. Juvenile retains yellow colouration until about 60cm in length, which may take 1–3 years. The change to green may take a few days or weeks; this process does not involve shedding skin.

NOT VENOMOUS but capable of inflicting painful and messy bite.

Australian Scrub Python *Simalia kinghorni*

LENGTH/ID: Australia's largest snake with reported maximum length of 8m. Average adult length 3.5m. Slender with long head. Scales smooth with iridescent sheen. Yellow-brown to brown with many irregular darker bands and blotches. Belly white to cream.

RANGE: From tip of eastern Cape York Peninsula along east coast to Townsville. Also Torres Strait islands.

HABITAT/HABITS: Rainforests, vine thickets, sclerophyll forest and humid woodland. Juveniles generally arboreal, adults mainly terrestrial but can climb. Nocturnal ambush predator feeding on mammals as large as wallabies.

NOT VENOMOUS but capable of inflicting painful and messy bite.

Oenpelli Rock Python *Simalia oenpelliensis*

LENGTH/ID: Average 3.5m, maximum 4.5m. Very slender relative to body length. Pale grey-brown to fawn with numerous brown to reddish-brown blotches that are largest down middle of back. Some of these blotches coalesce to form bars along body. Belly cream.

RANGE: Only in western Arnhem Land escarpment of NT.

HABITAT/HABITS: Woodlands and rock outcrops. Shelters in caves and crevices on the escarpment, and in tree hollows. Nocturnal. Preys on mammals and birds.

NOT VENOMOUS but capable of inflicting painful and messy bite. Regarded as a docile snake that rarely bites.

File Snakes   Family Acrochordidae

There are three species in the file snake family, two of which occur in Australia although they are also found elsewhere. Skin covered in keels which make it quite rasp-like to touch, hence the name file snakes. These rough scales enable the snake to hold slippery fish by coiling around them. The head is blunt and the nostrils placed well forward to allow the snake to breathe with just the tip of its snout above water. They are totally aquatic and move very gracefully in the water, but on land they are almost helpless and ungainly.

Arafura File Snake *Acrochordus arafurae*

LENGTH/ID: Average 1.5m, maximum 2m. Stout-bodied with loose baggy grey to brown skin. Dark band down middle of back extends down sides to belly and encloses paler blotches.

RANGE: Northern Australia from NT to Cape York Peninsula, Qld.

HABITAT/HABITS: Totally aquatic. Eats fish. Inhabits freshwater lagoons and rivers. During wet season disperses widely across wetlands. A live-bearer that gives birth to 11–25 young. Nocturnal. In day shelters under overhanging banks or among submerged roots. A prized food for the Aboriginal people.

NOT VENOMOUS and completely harmless.

Little File Snake *Acrochordus granulatus*

LENGTH/ID: Average 80cm, maximum 1.5m. Blackish-grey to dark brown with numerous narrow pale bands. Belly colour similar but paler. Head blunt with small eyes and nostrils well forward.

RANGE: Northern Australia from Kimberleys in WA, across NT to Cape York Peninsula in Qld.

HABITAT/HABITS: Saltwater or brackish mud flats, mangroves and reef flats. Sometimes coastal freshwater areas. Mainly nocturnal; hides by day among mangrove roots or buried in debris. Live-bearing with 1–12 young. Fish-eater, will also take crabs.

NOT VENOMOUS and completely harmless.

Colubrid Snakes Family Colubridae

This family contains more than half the world's snakes, with about 1,600 species. They are found on all continents except Antarctica and also on many of the larger islands. Tends to be the predominant snake family in most areas where they occur, but this is not the case in Australia where the venomous elapid snakes occupy most habitats. The six colubrid species found here are recent arrivals geologically speaking and, having originated in New Guinea, are confined to northern and eastern Australia. None are found in the arid interior and they all utilise tropical and subtropical terrestrial habitats and waterways.

Brown Tree Snake or Night Tiger *Boiga irregularis*

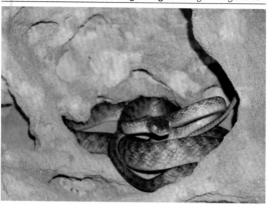

LENGTH/ID: Average 1.3m, maximum 1.7m. Slender with large broad head that is very distinct from neck. Large prominent eyes with vertical pupils. Eastern form pale brown to reddish-brown with narrow, irregular darker bands. Belly orange to salmon-coloured. Night Tiger form is cream with prominent red bands across body. Belly creamish.

RANGE: Eastern form along east coast from Sydney, NSW, to Cape York Peninsula, Qld. Night Tiger form occurs west of Cape York Peninsula across north of NT to Kimberley, WA.

Night Tiger.

HABITAT/HABITS: Arboreal and nocturnal, the Brown Tree Snake is found in a range of habitats including rock outcrops and escarpments, rainforest and other forests. Feeds on small mammals, birds and bird eggs, lizards and frogs. Sometimes found in birdcages where, having eaten the occupant, it cannot get back out through bars. Egg-laying species with 3–11 eggs per clutch. Aggressive when disturbed it will raise the fore body into a series of S-shaped loops and strike repeatedly.

WEAKLY VENOMOUS but not regarded as dangerous. However a bite from a large individual should be treated with caution.

Northern Tree Snake *Dendrelaphis calligastra*

LENGTH/ID: Average 80cm, maximum 1.2m. Brown to greenish-olive with black stripe running from snout through eye onto fore body. Very slender with very large eye.

RANGE: Far north-eastern Qld from about Paluma to eastern Cape York Peninsula.

HABITAT/HABITS: Dense vegetation along creeks, rainforest, vine thickets and urban areas. Arboreal, diurnal and egg-laying with 5–11 eggs per clutch. A very alert and fast; sometimes glimpsed coiled among foliage before making a speedy exit. Eats frogs and lizards.

NOT VENOMOUS and completely harmless.

Common Tree Snake or Green Tree Snake
Dendrelaphis punctulatus

LENGTH/ID: Average 1m, maximum 2m. Very slender with long thin tail and large prominent eyes. Colour extremely variable and seems to depend on geographic region in which individual occurs. In NSW and south-eastern Qld grey-green to olive-green with bright yellow belly. Can be brown to black with yellow belly in mid-east and north-east coast of Qld, and yellow with cream belly and blue-grey head and neck across northern Australia and WA. Occasionally a blue form is seen on east coast.

A ridge or keel along edges of belly assists the Common Tree Snake in maintaining a grip when climbing. When agitated inflates body and blue skin between scales is displayed.

RANGE: From Kimberley region, WA, across NT and Qld to Cape York Peninsula and down east coast to just south of Sydney, NSW.

HABITAT/HABITS: Widespread in well-vegetated areas including rainforest edges, coastal heath, woodlands and suburban gardens. Diurnal and arboreal.

Fast-moving with very good eyesight. Often forages for frogs and lizards on ground or in low vegetation. Has been observed taking tadpoles and fish from shallow water. If handled can give off an offensive odour. Sometimes colonies of this snake may be found in the roofs of houses, hollow trees and rock crevices where they aggregate in winter. Egg-laying with 3–16 eggs to a clutch.

NOT VENOMOUS and reluctant to bite.

Slatey-grey Snake *Stegonotus cucullatus*

LENGTH/ID: Average 90cm, maximum 1.5m. Uniform dark brown to black. Very glossy with iridescent sheen. White belly usually with some dark flecks. Eyes small and black.

RANGE: Northern parts of NT and Qld south to around Townsville.

HABITAT/HABITS: Mainly riverine habitats or adjacent moist areas. Nocturnal. Terrestrial, but also in low shrubs and water. Feeds on frogs, lizards, lizard eggs and occasionally mammals. Egg-laying with 7–16 eggs in a clutch. Emits pungent unpleasant odour if handled.

NOT VENOMOUS and harmless, but aggressive and strikes repeatedly if provoked.

Keelback or Freshwater Snake
Tropidonophis mairii

LENGTH/ID: Average 50cm, maximum 90cm. As common name suggests has strongly keeled scales. Colour variable – grey, black, brown, olive, reddish-brown or yellow. Often there are darker blotches or broken crossbands. Belly can be cream, olive-brown or salmon-coloured.

RANGE: Across northern Australia from Kimberley region, WA, through NT across to Cape York Peninsula, Qld, and down east coast to Clarence River area, NSW.

HABITAT/HABITS: The Keelback is semi-aquatic and can occur in large numbers around creeks, rivers, lagoons or swamps. Mainly nocturnal but occasionally seen during day. Food consists of frogs, tadpoles, fish and small lizards. Able to eat young Cane Toads without any ill effect, but does not seem to have same immunity with adults. This snake is unusual in that it can discard its tail if this is grasped by a predator. If handled roughly it will emit an unpleasant odour.

NOT VENOMOUS and harmless, but it will bite readily if provoked.

Mangrove and Freshwater Snakes
Family Homalopsidae

Only five of these aquatic or semi-aquatic snakes are found in Australia, although about 40 species occur worldwide. They have a tail that is round and tapering, as opposed to flattened and paddle-shaped in sea snakes. Adaptations to their aquatic lifestyle include nostrils located on top of the snout and a valve that shuts when the snake is underwater. The eyes are also directed upwards, meaning they can see and breathe with only a small portion of the head above water. All are venomous with fangs at the back of the mouth, but they are reluctant to bite and not considered dangerous to humans.

Australian Bockadam *Cerberus australis*

LENGTH/ID: Average 60cm, maximum 1.2m. Scales keeled and grey to brick-red with narrow, irregular black bands or streaks that are sometimes broken. Has dark streak through small, upward-facing eyes. Nostrils are on upper surface of snout. Belly cream to salmon-coloured with black bands.

RANGE: Northern WA, northern NT and north-west Qld to Cape York Peninsula.

HABITAT/HABITS: Aquatic and rarely leaves water. Estuaries, tidal creeks and rivers in brackish or saltwater, sheltering in mangroves or debris. Hunts small fish at night.

WEAKLY VENOMOUS but not regarded as dangerous.

White-bellied Mangrove Snake
Fordonia leucobalia

LENGTH/ID: Average 60cm, maximum 1m. Has smooth scales, a short blunt head and small eyes. Extremely variable in colour ranging from a glossy dark brown to black, red or yellow. There may or may not be numerous transverse white, yellow or red blotches or bands. Belly cream with a black stripe under tail.

RANGE: Along coastline of northern WA, NT and north-western Qld to tip of Cape York Peninsula and islands in Torres Strait.

HABITAT/HABITS: A nocturnal and aquatic species found in intertidal channels in mangrove swamps and on mud flats where it shelters in crab holes or in the mud. Feeds mainly on crabs and is unique among Australian snakes in dismembering larger crabs before eating them. Will strike above the crab then press it down into the mud, then pull off the legs and claws one at a time and eat them. Finally the crab's body is swallowed whole.

WEAKLY VENOMOUS, but not regarded as dangerous.

Macleay's Water Snake *Pseudoferania polylepis*

LENGTH/ID: Average 60cm, maximum 1m. Smooth-scaled and glossy dark brown to blackish, often with darker vertebral stripe and another along lower side. Belly cream to yellow with black stripe under tail.

RANGE: Northern NT and Qld to Cape York Peninsula.

HABITAT/HABITS: Freshwater lagoons, swamps and water courses where it shelters in submerged roots, overhanging banks and aquatic vegetation. Usually nocturnal but sometimes seen hunting fish and frogs during day.

WEAKLY VENOMOUS but not regarded as dangerous.

Venomous Land Snakes   Family Elapidae
Subfamily Elapinae

The venomous land snakes or elapids make up the majority of land snakes in Australia, unlike most other continents where the colubrids are the dominant family. Of about 178 Australian snake species, 106 (60 per cent) are elapids. However most are not dangerous to humans.

They are front-fanged with the enlarged hollow teeth at the front of the upper jaw connected to a venom gland at the back of the head. All are predators, using their venom to kill or immobilise their prey before eating it whole.

The elapids are characterised by having enlarged and symmetrically arranged scales on the top of the head. The belly scales are also enlarged. These snakes come in all shapes and sizes ranging from 30cm to giants up to 3m long. Some are slender and fast-moving while others are bulky and quite sedentary. Almost all are terrestrial and while many can climb trees only a few are regarded as being arboreal.

Common Death Adder *Acanthophis antarcticus*

LENGTH/ID: Average 60cm, maximum 1m. A bulky solid snake with a broad triangular head and a thick body ending abruptly at a thin tail. Red-brown to grey in colour with irregular bands across body. Head shields slightly wrinkled while scales on body are smooth or very weakly keeled.

RANGE: Southern WA and SA, central and eastern NSW and Qld, and eastern NT.

HABITAT/HABITS: Heaths, shrublands and woodlands, often in rocky areas. Prefers undisturbed habitat and has declined in many areas because of land clearance, stock grazing and too frequent burning. Attracts lizards, birds or small mammals by concealing itself in loose soil or litter with tail in front of head. Tail tip is wriggled to attract prey.

DANGEROUSLY VENOMOUS, always capable of inflicting a fatal bite.

Kimberley Death Adder *Acanthophis lancasteri*

LENGTH/ID: Average 55cm, maximum 70cm. Bulky solid snake with broad triangular head and thick body ending abruptly at thin black-tipped tail. Dull orange, tan to grey above with numerous darker bands. Lips pale with darker mottling. Head shields wrinkled, and scales on fore part of body prominently keeled. Belly cream, darker on sides.

RANGE: Confined to Kimberley region of WA.

HABITAT/HABITS: This short, thick, well-camouflaged snake is most active at night, feeding on lizards and small mammals.

DANGEROUSLY VENOMOUS, always capable of inflicting a fatal bite.

Northern Death Adder *Acanthophis praelongus*

LENGTH/ID: Average 50cm, maximum 70cm. Bulky solid snake with broad triangular head and thick body ending abruptly at thin tail. Grey to reddish-brown with weak to prominent crossbands. Head usually darker with barred lips. Head shields moderately to strongly wrinkled and scales on fore part of body strongly keeled.

RANGE: North-western Qld and Cape York Peninsula.

HABITAT/HABITS: Grasslands and woodlands particularly around rocky outcrops. As with other death adders lies concealed in leaf litter or loose soil to lure prey.

DANGEROUSLY VENOMOUS, always capable of inflicting a fatal bite.

Desert Death Adder *Acanthophis pyrrhus*

LENGTH/ID: Average 50cm, maximum 70cm. Bulky with broad triangular head and thick body ending abruptly at thin tail. Red to red-brown with numerous distinct yellowish bands. Head shields very strongly wrinkled and body scales strongly keeled.

RANGE: Central Australia from western Qld to coastal WA.

HABITAT/HABITS: Spinifex deserts including both sandy and rocky areas and rock outcrops. Often concealed under edges of spinifex clumps waiting for prey to pass; almost impossible to detect in this situation.

DANGEROUSLY VENOMOUS, always capable of inflicting fatal bite.

Pilbara Death Adder *Acanthophis wellsei*

LENGTH/ID: Average 50cm, maximum 70cm. Small and relatively slender with a broad triangular head and body ending abruptly at thin tail. Red-brown with either pale brown or black bands. Those with black bands also have a black head. Head shields slightly wrinkled and body scales moderately keeled.

RANGE: Pilbara region and North West Cape in WA.

HABITAT/HABITS: Spinifex areas including both rocky areas and rock outcrops. Will often lie concealed under the edges of spinifex clumps.

DANGEROUSLY VENOMOUS, always capable of inflicting a fatal bite.

Pygmy Copperhead *Austrelaps labialis*

LENGTH/ID: Average 80cm, maximum 1.2m. Pale to dark grey to black, usually with dark bar across neck. Lips prominently barred and belly cream to dark grey.

RANGE: Southern Mount Lofty Ranges and Fleurieu Peninsula in SA and also on Kangaroo Island.

HABITAT/HABITS: On mainland generally found in higher altitude forests that have dense understorey. On Kangaroo Island occupies a wide range of habitats from coastal dunes to woodlands. Feeds mainly on frogs.

DANGEROUSLY VENOMOUS, capable of inflicting a fatal bite.

Highlands Copperhead *Austrelaps ramsayi*

LENGTH/ID: Average 1m, maximum 1.7m. Varies from dark grey to reddish-brown or black above with prominently barred lips. Lower sides usually paler or orange to red.

RANGE: Upland regions of eastern NSW to eastern Vic.

HABITAT/HABITS: Moist habitats in high altitude areas of grassland or forest where can occur in small aggregations. Shelters in hollow logs, animal burrows or rock piles. Very tolerant of cold. Feeds on lizards and frogs.

DANGEROUSLY
VENOMOUS. Always
capable of inflicting a fatal
bite, but reluctant to bite.

Juvenile.

Lowlands Copperhead *Austrelaps superbus*

LENGTH/ID: Average 1.2m; maximum 1.8m; largest individuals on King Island in Bass Strait. Ranges from pale brown to reddish-brown to almost black with no bands or markings. Head paler than body, lips only weakly barred. Lower sides usually paler or orange to red.

RANGE: Tas, some of Bass Strait islands, southern Vic from NSW border to south-eastern SA.

HABITAT/HABITS: Tussock grass areas in swampy areas or woodlands. Shelters in thick grass clumps.

DANGEROUSLY VENOMOUS. Always capable of inflicting a fatal bite, but reluctant to bite.

Australian Coral Snake *Brachyurophis australis*

LENGTH/ID: Average 30cm, maximum 45cm. Pink to red above with broad black band across head and another on nape. Numerous narrow irregular cream crossbands with dark edges to scales.

RANGE: From eastern SA and northern Vic through interior of NSW and Qld to Townsville area. Found by coast in northern NSW and southern Qld.

Adult and juvenile.

HABITAT/HABITS: Open woodlands, mallee scrub and grasslands on sandy soils. A secretive burrowing species which is usually seen on surface of ground at night. Feeds on lizards and reptile eggs. An egg-layer with 4–6 eggs in a clutch.

VENOMOUS but regarded as virtually harmless. Very reluctant to bite.

Narrow-banded Shovel-nosed Snake
Brachyurophis fasciolatus

LENGTH/ID: Average 30cm, maximum 40cm. Cream to pink above with broad black band across head and nape. Numerous narrow crossbands along body formed by black-tipped scales.

RANGE: From south-western WA through arid SA to north-western Qld and NSW.

HABITAT/HABITS: Grasslands and shrublands in arid sandy areas. Nocturnal burrowing species that feeds on lizards and reptile eggs.

VENOMOUS but regarded as virtually harmless. Very reluctant to bite.

Campbell's Shovel-nosed Snake
Brachyurophis campbelli

LENGTH/ID: Average 30cm, maximum 40cm. Fawn or orange to brown with broad black band across head. Numerous dark brown bands on body.

RANGE: Endemic to far northern Qld, from Mount Isa through to Cape York Peninsula.

HABITAT/HABITS: Nocturnal burrowing snake found in a variety of tropical woodlands and grasslands. A specialist feeder that probably feeds mainly on reptile eggs.

VENOMOUS but regarded as virtually harmless. Very reluctant to bite.

White-crowned Snake *Cacophis harriettae*

LENGTH/ID: Average 35cm, maximum 55cm. Grey to dark brown above with broad white to cream band across nape, narrowing along side of head and enclosing snout. Top of head black. Dark grey underneath.

RANGE: East coast and adjacent areas from northern NSW to Townsville area of Qld.

HABITAT/HABITS: Nocturnal. Shelters under logs and in deep leaf litter in damper areas. Also found in suburban areas in well-watered gardens and compost heaps.

VENOMOUS but regarded as virtually harmless.

Southern Dwarf Crowned Snake *Cacophis krefftii*

LENGTH/ID: Average 25cm, maximum 35cm. Black to dark brown above with narrow pale band around head, forming collar across nape. Creamy-yellow underneath with bands formed by black-edged scales.

RANGE: East coast from south-eastern Qld to central coast of NSW.

HABITAT/HABITS: Secretive, nocturnal and terrestrial. Hunts sleeping lizards. Shelters under logs and leaf litter in rainforests and other damp areas. When provoked will raise front of body off ground and point head downwards to display head pattern.

VENOMOUS but regarded as virtually harmless.

Golden-crowned Snake *Cacophis squamulosus*

LENGTH/ID: Average 45cm, maximum 75cm. Dark brown to dark grey above with paler stripe across snout and along face but not joining across nape. Pink to orange underneath with black spots along midline.

RANGE: East coast and adjacent ranges from central NSW to south-eastern Qld.

HABITAT/HABITS: Rainforest and damper areas of other forests. Terrestrial and nocturnal. When disturbed gives impressive display with fore body raised high off ground, but very reluctant to bite.

VENOMOUS. Large specimen may cause marked local symptoms.

Carpentaria Snake *Cryptophis boschmai*

LENGTH/ID: Average 40cm, maximum 55cm. Glossy and smooth scales. Pale brown to dark orange-brown above, paler on side of head and neck. Belly cream.

RANGE: Eastern Qld north to Cape York Peninsula.

HABITAT/HABITS: Woodlands and drier forests. Shelters beneath rocks, logs or debris. Terrestrial and nocturnal. Lizards and frogs are main food items.

VENOMOUS but not regarded as dangerous

Eastern Small-eyed Snake *Cryptophis nigrescens*

LENGTH/ID: Average 50cm, maximum 1m. Shiny black above without any pattern. Eyes small and black. Belly white to pinkish, often with black flecks.

RANGE: East coast and ranges from southern Vic, through NSW to Cairns district in Qld.

HABITAT/HABITS: Wide variety of habitats from rainforest to heathlands and rock outcrops. Terrestrial and nocturnal. Shelters under rocks and logs. Often found under bark on fallen trees.

DANGEROUSLY VENOMOUS, capable of inflicting a fatal bite.

Black-striped Snake *Cryptophis nigrostriatus*

LENGTH/ID: Average 40cm, maximum 55m. A slender snake. Glossy dark brown to reddish-pink above with black to dark brown head and broad dark vertebral stripe to tail tip. Cream to white underneath.

RANGE: Qld from Rockhampton north to Torres Strait islands.

HABITAT/HABITS: Terrestrial and nocturnal, this secretive snake is found in dry woodlands where it shelters beneath logs and rocks. Feeds mainly on lizards.

VENOMOUS but not regarded as dangerous.

Narrow-headed Whipsnake *Demansia angusticeps*

LENGTH/ID: Average 85cm, maximum 1m. Olive-brown to greyish-brown above with large prominent eyes. Dark line across snout and pale bar in front of and behind eye, with dark brown comma below eye. Underneath cream to yellow with grey spots on throat and fore body.

RANGE: Southern Kimberley region of WA to Victoria River area of NT.

HABITAT/HABITS: Diurnal and fast-moving this whipsnake occurs in dry open habitat in tropical woodlands.

VENOMOUS but not regarded as dangerous.

Greater Black Whipsnake *Demansia papuensis*

LENGTH/ID: Average 1m, maximum 1.6m. Dark grey to black above. Head paler with scattered dark spots and eye often has narrow white rim. Grey below with whitish throat.

RANGE: Far northern Australia from Kimberley region in WA through NT to Cape York Peninsula in Qld.

HABITAT/HABITS: Tropical woodlands. A fast-moving, slender, diurnal snake that feeds on lizards. Despite Latin name not known to occur in New Guinea.

VENOMOUS and capable of causing marked local symptoms. Bites from large individuals are Potentially dangerous.

Yellow-faced Whipsnake *Demansia psammophis*

LENGTH/ID: Average 70cm, maximum 1m. Slender, fast-moving, diurnal snake. Grey to olive-brown above, often red-brown on neck and fore part of body. Has distinctive dark comma below eye and wide pale margin either side of eye. Belly grey-green to dull yellow.

RANGE: Wide distribution from mid-coastal and southern WA, north-western SA and adjacent NT, far north-west Vic extending into inland NSW to coast, and eastern Qld to Cape York Peninsula.

HABITAT/HABITS: Dry forests, open woodlands and heathlands extending into arid areas along watercourses. A lizard eater that utilises keen eyesight and speed to catch prey.

VENOMOUS and capable of causing marked local symptoms.

Blacksoil Whipsnake *Demansia rimicola*

LENGTH/ID: Average 60cm, maximum 1m. Grey to olive-brown above. Centres of scales darker than edges, producing narrow dark and light stripes along body. Has obscure pale-edged dark band across nape, pale-edged bar around front of snout, and comma-shaped mark below eye. Belly bright orange-red with lines of paired black spots on throat.

RANGE: Far north-east WA, through NT to interior of central Qld, north-western NSW and north-eastern SA.

HABITAT/HABITS: Grasslands and shrublands in arid areas. Shelters in grass clumps.

VENOMOUS and capable of causing marked local symptoms.

Collared Whipsnake *Demansia torquata*

LENGTH/ID: Average 50cm, maximum 85cm. A very slender whipsnake. Pale brown to grey-brown above with top of head darker. Has broad pale-edged dark bar across nape and narrow pale stripe from behind eye to nape. Blue-grey below.

RANGE: Coast and ranges in Qld from Gladstone to Cape York Peninsula.

HABITAT/HABITS: A swift-moving snake that is active during day and occurs across tropical woodlands and grasslands.

VENOMOUS but not regarded as dangerous.

Lesser Black Whipsnake *Demansia vestigiata*

LENGTH/ID: Average 1.2m, maximum 1.6m. A large whipsnake that is dark brown to dark grey above, often with a reddish tinge. Scales often darker-edged, giving reticulated appearance. No dark bar across front of snout. Yellow-grey below.

RANGE: From coastal south-eastern Qld north to Cape York and across far northern Australia to Kimberley region of WA.

HABITAT/HABITS: Drier habitats, particularly in woodlands and heathlands.

VENOMOUS and capable of causing marked local symptoms. Bites from large individuals are Potentially dangerous.

De Vis' Banded Snake *Denisonia devisi*

LENGTH/ID: Average 40cm, maximum 55cm. Small thickset snake with broad dark brown head. Pale yellow-brown above with narrow dark brown bands that may break up into blotches. Lips barred white and dark brown. Cream below.

RANGE: Western slopes and plains of central and south-east Qld and adjacent interior of northern NSW.

HABITAT/HABITS: Open woodlands and shrublands on cracking soils, where shelters under logs and other ground litter in moister areas. A frog-eater, it is also known as the 'mud adder'.

VENOMOUS, capable of causing marked local symptoms.

Ornamental Snake *Denisonia maculata*

LENGTH/ID: Average 40cm, maximum 50cm. Dark grey to brownish above with top of head darker and lips barred black and white. Belly cream-coloured.

RANGE: Endemic to Qld, occurring in central eastern regions.

HABITAT/HABITS: Very secretive, nocturnal and terrestrial. Occurs in low-lying cracking soil that is subject to flooding. Feeds on frogs. Can occur in high densities in suitable habitat.

VENOMOUS, capable of causing marked local symptoms. Bites from large individuals can be Potentially dangerous.

White-lipped Snake *Drysdalia coronoides*

LENGTH/ID: Average 35cm, maximum 45cm. Scales smooth, matt and grey-green to brown or black above. Has distinctive white streak edged above with black that runs from snout along upper lip to neck. Pink, orange or cream below.

RANGE: Found in Tas, eastern SA, coast and adjacent highlands of southern NSW and northern tablelands.

HABITAT/HABITS: Adapted to cool conditions, occurring in dry forests and subalpine woodlands with ground-cover of tussock grasses. Often seen basking on top of tussocks.

VENOMOUS but not regarded as dangerous.

Masters' Snake *Drysdalia mastersii*

LENGTH/ID: Average 30cm, maximum 40cm. Yellow-brown to grey above with darker head and pale band across nape. Has white stripe running along upper lip. Orange below with grey to black blotches along outer edge.

RANGE: Southern Australia from south-eastern WA to Big Desert region of Vic.

HABITAT/HABITS: Coastal dunes, heathlands and mallee spinifex habitat in sandy semi-arid areas.

VENOMOUS but not regarded as dangerous.

Mustard-bellied Snake *Drysdalia rhodogaster*

LENGTH/ID: Average 35cm, maximum 40cm. Brown to olive above with prominent pale brown to orange band across nape. Top of head dark brown, and has narrow black stripe from snout to eye. Yellow to orange below.

RANGE: Coast and ranges of NSW from Vic border to Blue Mountains.

HABITAT/HABITS: A terrestrial and diurnal snake found in dry forests, heathlands and tussock grasslands.

VENOMOUS but not regarded as dangerous.

Bardick *Echiopsis curta*

LENGTH/ID: Average 40cm, maximum 70cm. Short and thickset with broad head and smooth matt scales. Olive or brown to reddish-brown above, usually with pale flecks on head and white spots on lips. Greyish-brown below.

RANGE: One population in south-west WA, with another from Eyre Peninsula in south-east SA to adjacent Vic and south-western NSW.

HABITAT/HABITS: Mainly in mallee and spinifex areas. Terrestrial and nocturnal but may be seen basking during day at edge of spinifex or low bushes.

VENOMOUS, capable of causing marked local symptoms.

Red-naped Snake *Furina diadema*

LENGTH/ID: Average 30cm, maximum 40cm. Pale to dark red-brown above, each scale being darker-edged to give a reticulated appearance. Head and nape shiny black and enclose a bright red or orange patch. Creamy-white underneath.

RANGE: Interior and coast of NSW, eastern SA and eastern Qld to Cairns.

HABITAT/HABITS: Woodlands, dry forests and grasslands. Usually shelters beneath rocks but sometimes found inside termite nests.

VENOMOUS but not regarded as dangerous.

Dunmall's Snake *Furina dunmalli*

LENGTH/ID: Average 50cm, maximum 60cm. Pale to dark grey or brown above. A few scales on side of neck and fore body, and on upper lip, have yellow spots. Belly whitish.

RANGE: South-eastern interior of Qld and adjacent areas of NSW.

HABITAT/HABITS: Terrestrial and nocturnal. Occurs in Callitris and Acacia woodland in the Brigalow Belt. An uncommon species about which very little is known.

VENOMOUS, capable of causing marked local symptoms.

Moon Snake *Furina ornata*

LENGTH/ID: Average 45cm, maximum 65cm. Smooth and glossy slender-bodied snake. Yellow-brown to reddish-brown above, with darker-edged scales giving reticulated appearance. Head and nape glossy black, separated by red to orange band across nape. Cream below.

RANGE: Widespread across northern Australia but absent from southern and eastern areas.

HABITAT/HABITS: Wide range of habitats from tropical woodlands to hummock grasslands. Terrestrial and nocturnal.

VENOMOUS but not regarded as dangerous.

Grey Snake *Hemiaspis damelii*

LENGTH/ID: Average 50cm, maximum 65cm. Grey to olive-grey above, with black head that fades in older specimens to a black collar on neck. Belly cream with grey spots.

RANGE: Central inland NSW through south-eastern Qld to coast at about Rockhampton.

HABITAT/HABITS: Terrestrial and crepuscular to nocturnal, being most active in early evening. A frog-eater found in grasslands and open woodlands on floodplains where shelters in cracks in soil, in burrows or under debris.

VENOMOUS, capable of causing marked local symptoms.

Marsh Snake *Hemiaspis signata*

LENGTH/ID: Average 50cm, maximum 75cm. Olive to brown above with pale stripe from behind eye to neck and another pale stripe on upper lip. Black or dark grey below. Melanistic individuals occasionally encountered.

RANGE: Coast and ranges along east coast from around Nowra, NSW, to north-eastern Qld.

HABITAT/HABITS: Terrestrial and diurnal, can be nocturnal in warm weather. Wet forests and moist areas around swamps and creeks where sometimes found in small aggregations.

VENOMOUS, capable of causing marked local symptoms.

Pale-headed Snake *Hoplocephalus bitorquatus*

LENGTH/ID: Average 60cm, maximum 90cm. Pale grey to brown above with cream band on neck. Head pale grey with some dark grey to black scales. Lips barred. Belly cream, sometimes with darker spots.

RANGE: Patchily distributed along ranges and western slopes from central coast of NSW to Cape York Peninsula, Qld.

HABITAT/HABITS: Nocturnal and arboreal. Dry forest to rainforest. Usually near watercourses where shelters under loose bark of trees or in hollow limbs.

VENOMOUS, capable of causing marked local symptoms.

Broad-headed Snake *Hoplocephalus bungaroides*

LENGTH/ID: Average 60cm, maximum 90cm. Black above with yellow spots that form narrow irregular crossbands. Head broad and flattened. Belly grey.

RANGE: NSW coast and tablelands centred around Sydney, extending south to Nowra and west to Blue Mountains.

HABITAT/HABITS: Sandstone ridges in forested areas where shelters beneath rock slabs and in tree hollows. Nocturnal. Adopts very aggressive pose if disturbed. Because of similarity in colour pattern has been mistaken for the harmless Diamond Python.

VENOMOUS, capable of causing marked local symptoms. Potentially dangerous.

Stephens' Banded Snake *Hoplocephalus stephensii*

LENGTH/ID: Average 70cm, maximum 1m. Pale brown to yellow with series of broad black crossbands on body and tail. Some individuals lack crossbands. Head black with brown patch on top, white bars on lips and pale patch behind eye and on nape. Cream below with dark spots.

RANGE: Coastal ranges from central coast of NSW to south-eastern Qld.

HABITAT/HABITS: Wet forest including rainforest. Nocturnal and arboreal. Shelters in hollow tree limbs and rock crevices.

VENOMOUS, capable of causing marked local symptoms. Potentially dangerous.

Black-striped Burrowing Snake *Neelaps calonotos*

LENGTH/ID: Average 20cm, maximum 28cm. Small and slender with smooth and shiny scales. Bright orange-red above with cream centre to each scale. Has black vertebral stripe with cream spots from nape to tail. Black band extends across nape with another across top of head.

RANGE: Restricted to sandy coastal region around Perth, WA. Urban expansion an increasing threat.

HABITAT/HABITS: Coastal heaths and shrublands along dunes and sandplains. A burrowing species but active above ground at night. Lizard-eater specialising in very small burrowing skinks.

VENOMOUS but regarded as virtually harmless.

Tiger Snake *Notechis scutatus*

LENGTH/ID: Average 1m, maximum 2m. Large and robust with broad, blunt head. Huge variation in both colour and pattern, particularly in Tas. Colours range from silver-grey and yellow to olive and black, with bands that may be broad or narrow, prominent or obscure. Some individuals lack bands completely. Belly creamy yellow to grey, darker grey towards tail.

RANGE: South-western WA, south-eastern SA, Tas and Bass Strait islands, most of Vic excluding north-west, eastern NSW and southern Qld.

HABITAT/HABITS: Favours cooler more moist areas around swamps, lakes and rivers in grasslands and forested areas. Feeds on frogs, small mammals and birds. On Chappell Island in Bass Strait adults feed on muttonbird (shearwater) chicks.

DANGEROUSLY VENOMOUS, always capable of inflicting a fatal bite.

A Tasmanian Tiger Snake

A Chappell Island Tiger Snake.

Tiger Snake from WA.

Inland Taipan *Oxyuranus microlepidotus*

LENGTH/ID: Average 1.7m, maximum 2.5m. Pale brown to dark brown above with darker brown or black edging to scales which may form indistinct bands or a reticulated pattern. Head glossy black during winter but becomes paler during summer. Creamy-yellow below.

RANGE: South-western Qld and north-eastern SA. Known historically from far western NSW.

HABITAT/HABITS: Inhabits black soil plains and floodplains where it shelters in extensive soil cracks and burrows of Long-haired Rat. Although it has the most potent venom of any terrestrial snake it is secretive and rarely encountered. Sometimes called the 'fierce snake' – a name which is quite unjustified.

DANGEROUSLY VENOMOUS, always capable of inflicting a fatal bite.

Taipan *Oxyuranus scutellatus*

LENGTH/ID:
Average 1.8m,
maximum 3m.
Pale brown to
dark brown
above. Long
rectangular head
often pale cream
but sometimes
lighter colour
limited to snout.
Eye reddish.
Brow angular
and overhangs eye. Cream below, usually with orange spots.

RANGE: Extreme north-east NSW through coastal eastern Qld.
Also northern areas of NT into Kimberley region of WA.

HABITAT/HABITS: Diurnal, mainly active early morning and late
afternoon. Canefields, grassy woodlands and forests, sheltering in
logs or abandoned animal burrows. Feeds exclusively on mammals.

DANGEROUSLY VENOMOUS,
always capable of inflicting a
fatal bite.

130

Western Desert Taipan *Oxyuranus temporalis*

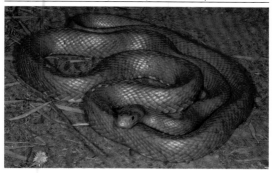

LENGTH/ID: Average 1.2m, maximum 1.7cm. Scales smooth and pale brown, sometimes with darker edges that form a reticulated pattern, especially on hind part of body. Head paler than rest of upper body. Belly creamy-yellow to grey.

RANGE: South-western NT and south-eastern WA. Only named in 2007 and known from three locations in western desert.

HABITAT/HABITS: Sandy areas with shrubby understorey dominated by spinifex under open woodland.

DANGEROUSLY VENOMOUS, always capable of inflicting a fatal bite.

Dwyer's Snake *Parasuta dwyeri*

LENGTH/ID: Average 40cm, maximum 60cm. Yellow-brown to orange-brown above with scales dark-edged. Occasionally has vague black vertebral stripe. Head and nape black with small pale patch in front of eye and on tip of snout. Lips and belly white.

RANGE: Central Vic and western slopes and ranges of NSW to south-eastern Qld.

HABITAT/HABITS: Woodlands, dry forests and shrublands. Terrestrial and nocturnal. Shelters beneath rocks and in cracks in soil.

VENOMOUS but regarded as virtually harmless.

Little Whip Snake *Parasuta flagellum*

LENGTH/ID: Average 30cm, maximum 45cm. Tan to orange above with darker edges to scales giving reticulated appearance. Top of head from eyes back to nape is black. From snout to eyes may be pale with no black band but usually has black band over snout. Belly pearly white.

RANGE: From south-eastern NSW to Vic and south-eastern SA.

HABITAT/HABITS: Open woodland, grasslands and rocky areas. Terrestrial and nocturnal. Shelters under rocks and often found in groups.

VENOMOUS but regarded as virtually harmless.

Gould's Hooded Snake *Parasuta gouldii*

LENGTH/ID: Average 40cm, maximum 53cm. Orange-brown to red-brown with each scale edged black to give reticulated appearance. Head black with pale blotch in front of each eye. Lower sides pale. Belly and lips white.

RANGE: South-western areas of WA.

HABITAT/HABITS: Heathlands and woodlands, often around rock outcrops. A nocturnal, lizard-eating snake.

VENOMOUS but regarded as virtually harmless.

Monk Snake *Parasuta monachus*

LENGTH/ID: Average 35cm, maximum 45cm. Brick-red with black head and nape. Scales sometimes edged in black to give reticulated appearance. Lips and belly white.

RANGE: Pilbara region of WA to eastern interior. Also south-western areas of NT and north-western SA.

HABITAT/HABITS: Occurs in semi-arid to arid regions, in woodlands, shrublands and rock outcrops. Secretive and nocturnal. Shelters under rocks and logs or in abandoned burrows.

VENOMOUS but regarded as virtually harmless.

Mitchell's Short-tailed Snake *Parasuta nigriceps*

LENGTH/ID: Average 35cm, maximum 59cm. Red-brown above with black head and neck. Has broad black vertebral stripe running from head to tail. Belly pearly white.

RANGE: Found in southern Australia from WA through SA to north-western Vic and adjacent areas of NSW.

HABITAT/HABITS: Semi-arid areas, particularly mallee woodland. Terrestrial and nocturnal, sheltering under logs, rocks or ground debris.

VENOMOUS but regarded as virtually harmless.

Mulga Snake *Pseudechis australis*

LENGTH/ID: Average 1.4m, maximum 2.5m. Large robust snake that is pale coppery-brown to red-brown or dark brown above. Sometimes the tip of each scale is dark-edged, resulting in a reticulated pattern. Cream to yellow below, sometimes with orange blotches.

RANGE: Widespread throughout Australia with exception of Tas, eastern NSW and south-eastern Qld, most of Vic, south-eastern SA and southern WA.

Juvenile.

HABITAT/HABITS: Occurs in many habitats from deserts to grasslands and woodlands. Terrestrial and diurnal but becomes nocturnal in hot weather, sheltering in logs, burrows or deep soil cracks. Also known as the King Brown Snake although it is a member of the black snake family.

DANGEROUSLY VENOMOUS, always capable of inflicting a fatal bite.

Spotted Mulga Snake *Pseudechis butleri*

LENGTH/ID: Average 1m, maximum 1.6m. Dark grey to black with yellow to brown spots on most scales. Side of head reddish-brown and rest of head and nape black. Cream to yellow underneath.

RANGE: Found only in arid mid-western interior of WA.

HABITAT/HABITS: Mulga woodlands and shrublands. Sometimes found in rocky areas. Diurnal but can become nocturnal in hot weather. Feeds on snakes, lizards, small mammals and frogs.

DANGEROUSLY VENOMOUS, capable of inflicting a fatal bite.

Collett's Snake *Pseudechis colletti*

LENGTH/ID: Average 1.2m, maximum 1.8m. Grey to reddish-brown or black with irregular cream to pinkish-orange bands or variegations across body. Lower sides are same colour as crossbands. Belly yellow to orange.

RANGE: Dry blacksoil plains of central Qld where it is also known as the Downs Tiger.

HABITAT/HABITS: Grasslands on cracking clay soils, where it shelters in cracks in soil. A diurnal, secretive snake that is not often seen.

DANGEROUSLY VENOMOUS, capable of inflicting a fatal bite.

Blue-bellied Black Snake *Pseudechis guttatus*

LENGTH/ID: Average 1.2m, maximum 1.5m. A large robust snake that is normally black above, sometimes with some scales blotched with cream. Belly grey to grey-blue.

There is a light-coloured variant that is cream with black edges to scales.

RANGE: South-eastern Qld through western slopes and plains of northern NSW to Hunter Valley.

HABITAT/HABITS: Terrestrial and diurnal, this black snake inhabits drier inland areas and river floodplains. Generally occurs in wooded habitat, sheltering in logs, burrows or ground debris. Also known as the Spotted Black Snake.

DANGEROUSLY VENOMOUS, capable of inflicting a fatal bite.

Juvenile.

142

Red-bellied Black Snake *Pseudechis porphyriacus*

LENGTH/ID: Average 1.2m, maximum 1.8m. Shiny black above with snout often pale grey or brown. Belly scales bright red at sides, becoming duller centrally. Adjacent body scales also often bright red.

RANGE: Eastern Australia from south-eastern SA, through Vic and NSW to far northern Qld.

HABITAT/HABITS: Diurnal. Frequents creeks, swamps and other water bodies. Feeds on frogs, small mammals, reptiles and fish. Populations of this snake have declined dramatically in areas where the introduced Cane Toad has become established.

DANGEROUSLY VENOMOUS, capable of inflicting a fatal bite.

Western Pygmy Mulga Snake *Pseudechis weigeli*

LENGTH/ID: Average 90cm, maximum 1.2m. Pale brown to dark reddish-brown, sometimes with a vague reticulated pattern formed by darker edges to scales. Head and nape sometimes have darker blotches or streaks. Cream below.

RANGE: Northern Australia from Kimberley region of WA to Mount Isa in Qld.

HABITAT/HABITS: Tropical woodlands to spinifex-dominated areas. The status of this snake is uncertain and it may in fact comprise more than one species.

DANGEROUSLY VENOMOUS, capable of inflicting a fatal bite.

Dugite *Pseudonaja affinis*

LENGTH/ID: Average 1m, maximum 2.1m. Large slender snake that is variable in colour. Can be olive, pale to dark brown, reddish-brown or black above, lightly to heavily spotted with black. Head may be paler or darker than body. Belly greyish-white to pale brown with orange blotches. The Rottnest Island form is blackish both above and below.

Rottnest Island Dugite.

RANGE: From south-western region of WA to Eyre Peninsula in SA. A dwarf form, *Pseudonaja affinis exilis*, occurs on Rottnest Island off Fremantle.

HABITAT/HABITS: Wide-ranging habitats from coastal dunes, woodlands and forests to disturbed rural and urban areas. Terrestrial and both diurnal and nocturnal depending on temperature. An alert snake that is quick to retreat but will bite if provoked.

DANGEROUSLY VENOMOUS, always capable of inflicting a fatal bite.

Strap-snouted Brown Snake
Pseudonaja aspidorhyncha

LENGTH/ID: Average 1m, maximum 1.3m. Highly variable in colour, from grey-brown to dark brown above with or without broad dark bands, or a reticulated pattern formed by dark edges to scales. Usually has some dark markings on nape that can range from a few dark scales to a broad band. Snout square and strap-like.

RANGE: Central NSW, south-western Qld, north-western Vic and eastern half of SA.

HABITAT/HABITS: Terrestrial and diurnal although becomes nocturnal in hot weather. Occurs in a wide range of habitats including deserts, woodlands and shrublands, as well as disturbed rural areas.

DANGEROUSLY VENOMOUS, always capable of inflicting a fatal bite.

Juvenile.

Speckled Brown Snake *Pseudonaja guttata*

LENGTH/ID: Average 80cm, maximum 1.2m. Slender medium-sized brown snake that is tan to light brown above with numerous black-edged scales. There is also a banded form with evenly spaced broad dark blotches from nape to tail. Sometimes has several narrow dark lines between bands. Belly cream with orange blotches.

RANGE: Barkly Tablelands of NT and central Qld to north-eastern corner of SA.

HABITAT/HABITS: Occurs in grasslands on blacksoil plains where it shelters in deep soil cracks. Feeds on small mammals, reptiles and frogs.

DANGEROUSLY VENOMOUS, always capable of inflicting a fatal bite.

Ingram's Brown Snake *Pseudonaja ingrami*

LENGTH/ID: Average 1.2m, maximum 1.8m. Large snake with narrow head. Head and neck brown to black and body ranges from pale to reddish-brown to dark brown with tip of scales being darker. Belly pale yellow with orange spots or orange-edged scales.

RANGE: Barkly Tableland of NT and adjacent areas of Qld interior. Another population known from Kununurra region in WA.

HABITAT/HABITS: Grasslands on cracking blacksoil plains, preferring more low-lying areas where it shelters in deep cracks in soil.

DANGEROUSLY VENOMOUS, always capable of inflicting a fatal bite.

Western Brown Snake or Gwardar
Pseudonaja mengdeni

LENGTH/ID: Average 1.1m, maximum 1.6m. Extremely variable in colour and pattern. Light brown to orange to dark brown, with or without black head and nape; black scales on nape either random in shape of 'V' or 'W'; broad dark bands with narrow lines in between; or variations of these patterns. Snout rounded when viewed from above. Cream to yellow below with orange blotches.

RANGE: Western and central Australia to western Qld and north-western NSW.

HABITAT/HABITS: Drier areas in a variety of habitats including spinifex, woodlands and agricultural areas. Shelters in burrows and logs or within grass clumps. This snake is also known under the common name of Gwardar.

DANGEROUSLY VENOMOUS, always capable of inflicting a fatal bite.

Ringed Brown Snake *Pseudonaja modesta*

LENGTH/ID: Average 45cm, maximum 60cm. Smallest of the brown snakes. Tan to red-brown above with evenly spaced narrow black bands across body and tail. Top of head black and has broad black band across nape. All these markings become indistinct with age. Cream to yellow below with orange blotches.

RANGE: Arid regions of all mainland states except Vic.

HABITAT/HABITS: Grasslands, woodlands, rocky outcrops and dry watercourses. Shelters in spinifex or ground debris.

VENOMOUS, capable of causing marked local symptoms.

Northern Brown Snake *Pseudonaja nuchalis*

LENGTH/ID: Average 1.1m, maximum 1.5m. Snout square and strap-like when viewed from above. Pale brown to very dark brown, usually with reticulated pattern formed by dark-edged scales. Snout may be paler than rest of head. Sometimes has broad dark bands across body.

RANGE: Tropical northern Australia from WA across NT to Qld.

HABITAT/HABITS: Savanna woodlands, rocky outcrops and ranges, also agricultural areas. Terrestrial and diurnal, also nocturnal in hot weather. Shelters in animal burrows, soil cracks or debris on ground.

DANGEROUSLY VENOMOUS, always capable of inflicting a fatal bite.

Common Brown Snake *Pseudonaja textilis*

LENGTH/ID: Average 1.3m, maximum 2.3m. Colour varies from pale tan to dark brown and almost black. Adults usually lack any markings but sometimes have lighter or darker flecking. Juveniles have black band across head and nape; some also have numerous narrow black bands on body. These patterns usually disappear with age. Belly cream to yellow with orange blotches.

RANGE: Widespread across Qld, NSW and Vic to south-eastern SA with isolated populations in central Australia and Barkly Tableland in NT.

HABITAT/HABITS: Widespread in grasslands, open woodlands and shrublands. Also in agricultural and urban areas. A very quick and alert snake that is terrestrial and diurnal.

DANGEROUSLY VENOMOUS, always capable of inflicting a fatal bite.

Juvenile.

Desert Banded Snake *Simoselaps anomalus*

LENGTH/ID: Average 15cm, maximum 25cm. Scales are smooth, glossy and yellow to orange-red with numerous black rings across body from nape to tail. Head black with white bar across head behind eyes and white patch on snout.

RANGE: North-west coast of WA through northern and central sandy deserts to south-western NT and north-western SA.

HABITAT/HABITS: Sandy spinifex-dominated arid areas. Shelters in loose sand under spinifex clumps or shrubs.

VENOMOUS but not regarded as dangerous.

Jan's Banded Snake *Simoselaps bertholdi*

LENGTH/ID: Average 20cm, maximum 30cm. Scales smooth, glossy and yellow to orange-red with each scale edged in orange. Has numerous black bands across body from nape to tail and a broad black band across neck. Head whitish with black flecks.

RANGE: Mid and south-western regions of WA extending east to western SA and southern NT.

HABITAT/HABITS: Arid sandy areas and dunes with grasses, heaths and woodlands. Shelters in loose soil and leaf litter beneath shrubs.

VENOMOUS but not regarded as dangerous

Rosen's Snake *Suta fasciata*

LENGTH/ID: Average 45cm, maximum 65cm. Smooth glossy scales that are olive-brown to orange-brown with numerous irregular dark bands or blotches across body. Head spotted and has dark line running from nostril through eye to side of neck.

RANGE: Endemic to WA where occurs on north-west coast and in south-western interior.

HABITAT/HABITS: Terrestrial and nocturnal. A lizard-eater found in shrublands and woodlands where it shelters beneath rocks and timber, soil cracks or old burrows.

VENOMOUS, capable of causing marked local symptoms.

Little Spotted Snake *Suta punctata*

LENGTH/ID: Average 40cm, maximum 55cm. Pale brown to reddish-brown smooth and glossy scales that are sometimes tipped black. Head usually paler with dark patches on top of head and along neck. Belly creamy-white below.

RANGE: Northern Australia from WA coast through NT to far western Qld.

HABITAT/HABITS: Nocturnal. Spinifex grasslands and tropical woodlands. Shelters beneath ground debris and in spinifex clumps.

VENOMOUS but not regarded as dangerous.

Curl Snake *Suta suta*

LENGTH/ID: Average 40cm, maximum 60cm. Pale brown to dark brown with scales sometimes darker-edged to give reticulated pattern. Head and nape darker brown with pale band from snout through eye and dark stripe above lips. Some individuals have broad black stripe down middle of back.

RANGE: Very widespread. NSW and Qld excluding coastal regions, central and eastern SA, NT excluding far north, and north-eastern WA.

HABITAT/HABITS: Woodlands, grasslands and shrublands in arid and semi-arid regions. Shelters in earth cracks, under logs and rocks and under ground debris.

VENOMOUS, capable of causing marked local symptoms.

Rough-scaled Snake *Tropidechis carinatus*

LENGTH/ID: Average 75cm, maximum 1m. Strongly keeled non-glossy scales. Olive-brown to dark brown above usually with narrow dark crossbands, more prominent on fore body. Belly cream to olive green.

RANGE: Coastal regions and adjacent ranges from central coast of NSW to south-eastern Qld, with isolated population between Townsville and Cooktown.

HABITAT/HABITS: Rainforests, wet forests and edges of streams and swamps. Nocturnal and semi-arboreal, sometimes basks on shrubs during day.

DANGEROUSLY VENOMOUS, capable of inflicting a fatal bite.

Common Bandy Bandy *Vermicella annulata*

LENGTH/ID: Average 55cm, maximum 85cm. Unlikely to be confused with any other snake in its range. Entire head, body and tail banded with distinct alternate black and white rings.

RANGE: Widespread distribution across eastern Australia including south-eastern SA, northern Vic, most of NSW, and eastern and central Qld, extending into NT.

Eating a blind snake.

HABITAT/HABITS: Nocturnal burrowing snake that is sometimes seen on surface at night, particularly after rain. Feeds mainly on blind snakes and occurs in habitats from forests, woodlands and shrublands to deserts. When threatened will thrash about and loop the body vertically.

VENOMOUS but regarded as harmless.

Centralian Bandy Bandy *Vermicella vermiformis*

LENGTH/ID: Average 50cm, maximum 80cm. Entire head and body banded with distinct alternate black and white rings that encircle whole body. Differs from Common Bandy Bandy in having a greater number of belly scales; the two species do not appear to occur together.

RANGE: Found in central Australia around Alice Springs, with another population in southern Arnhem Land.

HABITAT/HABITS: Woodlands, shrublands, rocky ranges and deserts. Nocturnal burrowing snake sometimes seen on surface at night, particularly during or after rain. Feeds mainly on blind snakes.

VENOMOUS but regarded as harmless.

SEA SNAKES Family Elapidae, subfamily Hydrophiinae

About 32 species of sea snakes occur in Australian waters and most of these are also found in New Guinea and Asia. The majority reside in tropical waters – those found in more southerly areas are vagrants and infrequent visitors. Sea snakes have vertically flattened, often paddle-shaped tails and their nostrils are in valves located on top of the head. All are expert swimmers and divers, with some able to dive to depths of 100m. All sea snakes give birth to live young at sea. They never come onto land unless washed up by storms.

SEA KRAITS Family Elapidae, subfamily Laticaudinae

The sea kraits are closely related to the terrestrial elapids. They are smooth-scaled with many black bands across the body. Unlike sea snakes they have laterally situated nostrils rather than on top of the snout. The ventral scales are broad and the tail paddle-shaped. They are not confined to water and are frequently found on land, often a considerable distance from the sea, but usually in rocky shorelines and crevices. Egg-laying with eggs laid on land in crevices or debris. They are mostly nocturnal but do bask on land during the day. No breeding colonies are known in Australia and any individuals found are waifs brought here by ocean currents.

Horned Sea Snake *Acalyptophis peronii*

LENGTH/ID: Average 1m, maximum 1.3m. Cream to pale brown, sometimes with broad dark crossbands. Has conspicuous projecting scales or 'horns' above and behind eyes. Fore body slender and becomes thicker towards tail.

RANGE: Coral reefs from mid north-west coast of WA to south-eastern Qld.

HABITAT/HABITS: Appears to be mostly nocturnal. Uses narrow head and fore body to poke into crevices and burrows for gobies.

DANGEROUSLY VENOMOUS, always capable of inflicting a fatal bite.

Dubois' Sea Snake *Aipysurus duboisii*

LENGTH/ID: Average 70cm, maximum 1.1m. Dark brown to black both above and on belly. Edges of scales are cream. Scales on sides of body often cream and form triangular patches. Scales on top of head are fragmented rather than symmetrically arranged.

RANGE: Widespread in northern Australian waters from mid north-west coast of WA to north coast of NSW.

HABITAT/HABITS: Found on coral reefs and reef flats in shallow water but can occur in depths of about 50m. Forages for fish on sea floor.

DANGEROUSLY VENOMOUS, always capable of inflicting a fatal bite.

Olive Sea Snake *Aipysurus laevis*

LENGTH/ID: Average 1.2m, maximum 1.7m. Brown to olive above with or without cream spots. Sometimes lower surfaces and tail are cream. Scales on top of head are enlarged but fragmented. Belly scales enlarged and each has small notch in middle of free edge.

RANGE: Northern coastal and reef waters from mid north-west coast of WA to central coast of NSW.

HABITAT/HABITS: Coral reefs. Sometimes found in very high numbers. Inquisitive, which has been interpreted by some divers as aggressive.

DANGEROUSLY VENOMOUS, always capable of inflicting a fatal bite.

Mosaic Sea Snake *Aipysurus mosaicus*

LENGTH/ID: Average 65cm, maximum 1m. Cream or orange to yellow-brown above with many scales being dark-edged. Head brown, sometimes flecked darker brown. Belly scales weakly keeled.

RANGE: Northern Australian waters from mid north-west coast of WA to north coast of NSW.

HABITAT/HABITS: Usually found in deeper waters or around estuaries. Feeds entirely on fish eggs and reluctant to bite even when provoked.

VENOMOUS, capable of causing marked local symptoms.

Stokes' Sea Snake *Astrotia stokesii*

LENGTH/ID: Average 1.2m, maximum 2m. Large bulky body and large head. Yellowish-brown to brown above, with or without darker blotches and narrow bands which become obscure in older individuals. Scales on top of head enlarged and regular. Belly scales small and aligned in pairs to form deep keel.

RANGE: Coasts from about Exmouth in WA across tropical north to south coast of NSW.

HABITAT/HABITS: Coral reefs and coastal waters. Forages for fish in crevices and holes. Usually seen on surface in shallower waters.

DANGEROUSLY VENOMOUS, always capable of inflicting a fatal bite.

Australian Beaked Sea Snake *Enhydrina zweifeli*

LENGTH/ID: Average 1.2m, maximum 1.5m. Robust grey body, sometimes with numerous darker crossbands. Head dark grey and body scales keeled. Mental shield very elongated and much longer than it is broad.

RANGE: Gulf of Carpentaria and west to north-east coast of WA.

HABITAT/HABITS: Inhabits estuarine areas rather than coral reefs. Not commonly encountered. Eats fish, including catfish. Aggressive when disturbed or handled.

DANGEROUSLY VENOMOUS, always capable of inflicting a fatal bite.

Elegant Sea Snake *Hydrophis elegans*

LENGTH/ID: Average 1.7m, maximum 2m. Large snake with slender neck and fore body becoming bulkier towards tail. Pale brown with dark head and numerous dark bands which are narrow or broken on side of body. This pattern most prominent on young snakes, becoming more obscure on adults.

RANGE: South-west coast of WA, across tropical north to central coast of NSW.

HABITAT/HABITS: Deeper reef and murky coastal waters. Feeds on eels, using thin head and fore body to extract them from crevices and holes.

DANGEROUSLY VENOMOUS, always capable of inflicting a fatal bite.

Spectacled Sea Snake *Hydrophis kingii*

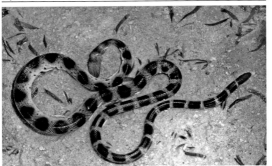

LENGTH/ID: Average 1.5m, maximum 1.9m. Fore part of body slender, becoming bulkier mid body. Grey above with black bands that become dark blotches towards tail, with lower side cream to grey. Head and throat black and has white ring around eye. Scales on top of head are enlarged and not fragmented. Belly scales small.

RANGE: From mid-western coast of WA across northern Australia to central coast of NSW.

HABITAT/HABITS: A fish-eater that is normally found in deeper waters.

DANGEROUSLY VENOMOUS, always capable of inflicting a fatal bite.

Ornate Sea Snake *Hydrophis ornatus*

LENGTH/ID: Average 1m, maximum 1.5m. Thickset with body uniformly wide along its length. Grey to bluish-grey above, with numerous darker bands or blotches, often obscure in adults. Usually has series of dark circular blotches along creamish lower sides.

RANGE: South-west coast of WA across northern Australia to central coast of NSW. Has even been recorded in Tas waters.

HABITAT/HABITS: Deeper waters of coral reefs, but also occurs in estuarine habitats.

DANGEROUSLY VENOMOUS, always capable of inflicting a fatal bite.

Yellow-bellied Sea Snake *Pelamis platura*

LENGTH/ID: Average 70cm, maximum 1m. Upper half of body black and lower half pale brown or yellow, with a sharp demarcation between the two colours. Tail yellow with black spots or bars.

RANGE: Widely distributed around Australian coastline except in far south.

HABITAT/HABITS: Pelagic species, widespread across world's oceans. Occurs in open seas rather than reefs or coastal waters, but often washed on to beaches by storms. Feed on surface fishes and capable of swimming backwards.

DANGEROUSLY VENOMOUS, always capable of inflicting a fatal bite.

Brown-lipped Sea Krait *Laticauda laticaudata*

LENGTH/ID: Average 80cm, maximum 1m. Blue to blue-grey above with numerous black bands across body. Top of head black with yellow bar above eyes that sometimes continues to snout. Upper lip brown and belly cream to yellow.

RANGE: Occurs rarely on tropical and temperate east coast of Australia.

HABITAT/HABITS: Shallow reefs along rocky shorelines and small islands that provide refuges ashore. Highly venomous but very inoffensive and reluctant to bite.

DANGEROUSLY VENOMOUS, always capable of inflicting a fatal bite, but quite docile and reluctant to bite.

FURTHER READING

Bush, B., Maryan, B., Browne-Cooper, R. and Robinson, D. 2007. *Reptiles and Frogs in the Bush: Southwestern Australia*. University of Western Australia Press.

Bush, B., Maryan, B., Browne-Cooper, R. and Robinson, D. 2010. *Field Guide to Reptiles and Frogs of the Perth Region*. WA Museum, Welshpool.

Cogger, H.G. 2014. *Reptiles and Amphibians of Australia*. CSIRO Publishing, Collingwood.

Couper, P. and Amey, A. 2007. *Snakes of South-east Queensland*. Queensland Museum, South Brisbane..

Coventry, A.J. and Robertson, P. 1991. *The Snakes of Victoria. A Guide to their Identification*. Department of Conservation and Environment, East Melbourne.

Emmott, A. and Wilson, S.G. 2009. *Snakes of Western Queensland, A Field Guide*. Desert Channels Qld, Longreach, Qld.

Fearn, S. 2015. *Snakes of Tasmania*. Queen Victoria Museum and Art Gallery, Tas.

Griffith, K. 2006. *Frogs and Reptiles of the Sydney Region*. Reed New Holland, Chatswood.

Heatwole, H. 1999. *Sea Snakes*. Australian Natural History Series, University of NSW Press, Sydney.

Johnson, S. 2015. *A Guide to Snakes of the NSW Mid North Coast*. Reptile Solutions, Frazers Creek, NSW.

McEwan, S. 2008. *Steve's Guide to Snakes of the East Coast and Tablelands of NSW.* Steve McEwan's Reptile World, NSW.

Michael, D. and Lindenmeyer, D. 2010. *Reptiles of the NSW Murray catchment.* CSIRO publishing, Collingwood.

Pearson, D. 2011. *Snakes of Western Australia.* Department of Environment and Conservation, Kensington, WA.

Shine, R. 2009. *Australian Snakes: A Natural History.* Reed New Holland, Chatswood.

Storr, G.M., Smith, L.A. and Johnstone, R.E. 2002. *Snakes of Western Australia.* Western Australian Museum, Perth.

Swan, G. and Sadlier, R., Shea, G. 2017. *A Field Guide to Reptiles of New South Wales.* Third Edition. Reed New Holland, Chatswood.

Swan, G. and Wilson, S. 2008. *What Snake is That? Introducing Australian Snakes.* Reed New Holland, Chatswood.

Swan, M. and Watherow, S. 2005. *Snakes, Lizards and Frogs of the Victorian Mallee.* CSIRO Publishing, Collingwood.

Swanson, S. 2007. *Field Guide to Australian Reptiles.* Steve Parish Publishing, Archerfield.

Wilson, S. 2015. *A Field Guide to Reptiles of Queensland.* Reed New Holland, Chatswood.

Wilson, S. and Swan, G. 2017. *A Complete Guide to Reptiles of Australia.* Fifth Edition. New Holland Publishers, Chatswood.

GLOSSARY

Anal: referring to anus or vent. Usually relates to scale covering vent.

Angular: in or at an angle.

Aquatic: living in or near water.

Arboreal: living in trees.

Brigalow: plant community dominated by acacias, usually *Acacia harpophylla*.

Callitris: cypress pine.

Crepuscular: active at dawn, dusk or in deeply shaded conditions.

Diurnal: active during daylight hours.

Egg tooth: forward-pointing tooth on snout of snakes and lizards. Used to cut through egg-shell at birth, falling off shortly after hatching.

Endemic: restricted to a particular region.

Escarpment: steep slope or cliff.

Estuarine: found in estuaries.

Hummock Grass: spinifex.

Iridescence: rainbow sheen on scales of some snakes which changes colour when angle of view is changed.

Keel: narrow raised ridge on individual scales.

Lateral: relating to the sides.

Longitudinal: running along the length.

Mallee: plant community dominated by mallee eucalypts which grow from lignotubers.

Matt: dull, non-shiny surface.

Melanistic: blackish, as a result of exceptional development of black pigment.

Mental shield: single midline scale on front edge of lower jaw.

Nape: the back of the neck.

Nocturnal: active after dark.

Parthenogenetic: able to reproduce without fertilisation by a male, particularly some geckos that produce female clones.

Pelagic: inhabiting open seas.

Reticulation: a net-like pattern. Usually refers to skin-patterns which contain a network of lines.

Riverine: relating to the bank of a river or other body of water.

Savanna: open grassland with scattered trees and bushes.

Sclerophyll forest: forest mostly comprising eucalypt (gum) trees which have hard stiff leaves.

Slough: the cast-off outermost skin-layer of a snake.

Spinifex: a common name applied to spiny-leafed grasses of the genus *Triodia*, which form prickly hummocks. Often called porcupine grass.

Spur: a rigid spine-like structure on either side of the vent, as in the vestigial hind-limbs of pythons.

Terrestrial: living on the ground.

Trilobed: having three lobes, as in the snouts of some blind snakes.

Transverse: crosswise, across the body.

Tubercles: a rounded or pointed projection.

Vent: transverse external opening of the anus.

Ventral: the lower (under) surfaces or the scales of the belly.

Vertebrate: an animal having a backbone or vertebral column.

INDEX

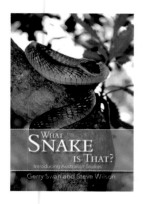 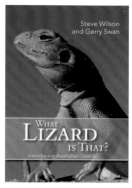

TITLES IN THE WHAT ANIMAL IS THAT? SERIES

What Snake is That? Introducing Australian Snakes
by Gerry Swan and Steve Wilson (ISBN 9781877069574)

What Lizard is That? Introducing Australian Lizards
by Steve Wilson and Gerry Swan (ISBN 9781877069581)

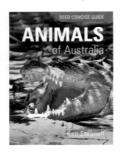

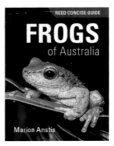

OTHER TITLES IN THE REED CONCISE GUIDES SERIES

Animals of Australia (ISBN 978 1 92151 754 9)
Wildflowers of Australia (ISBN 978 1 92151 755 6)
Birds of Australia (ISBN 978 1 92151 753 2)
Frogs of Australia (ISBN 978 1 92151 790 7)